Lean Manufacturing

BY Daniel LUCKY

Table of Content

SCIENTIFIC MANAGEMENT

MASS PRODUCTION TACTICS

LEAN TECHNIQUES

MANUFACTURING TECHNIQUES

GROUP TECHNOLOGY / CELLULAR MANUFACTURING,

SET-UP TIME REDUCTION

 SET-UP ANALYSIS

OVERALL EQUIPMENT EFFECTIVENESS

SMALL MACHINE CONCEPT

LEAN MATERIAL CONTROL

"PULL" SYSTEMS

LEAN PURCHASING

ORGANIZATION FOR CHANGE

TOTAL QUALITY MANAGEMENT

THE DEMING CYCLE

LEAN FLEXIBILITY

REDUCING UNCERTAINTY

TOTAL PREVENTATIVE MAINTENANCE

THE PROBLEM WITH INVENTORY

OPERATIONAL PREREQUISITE

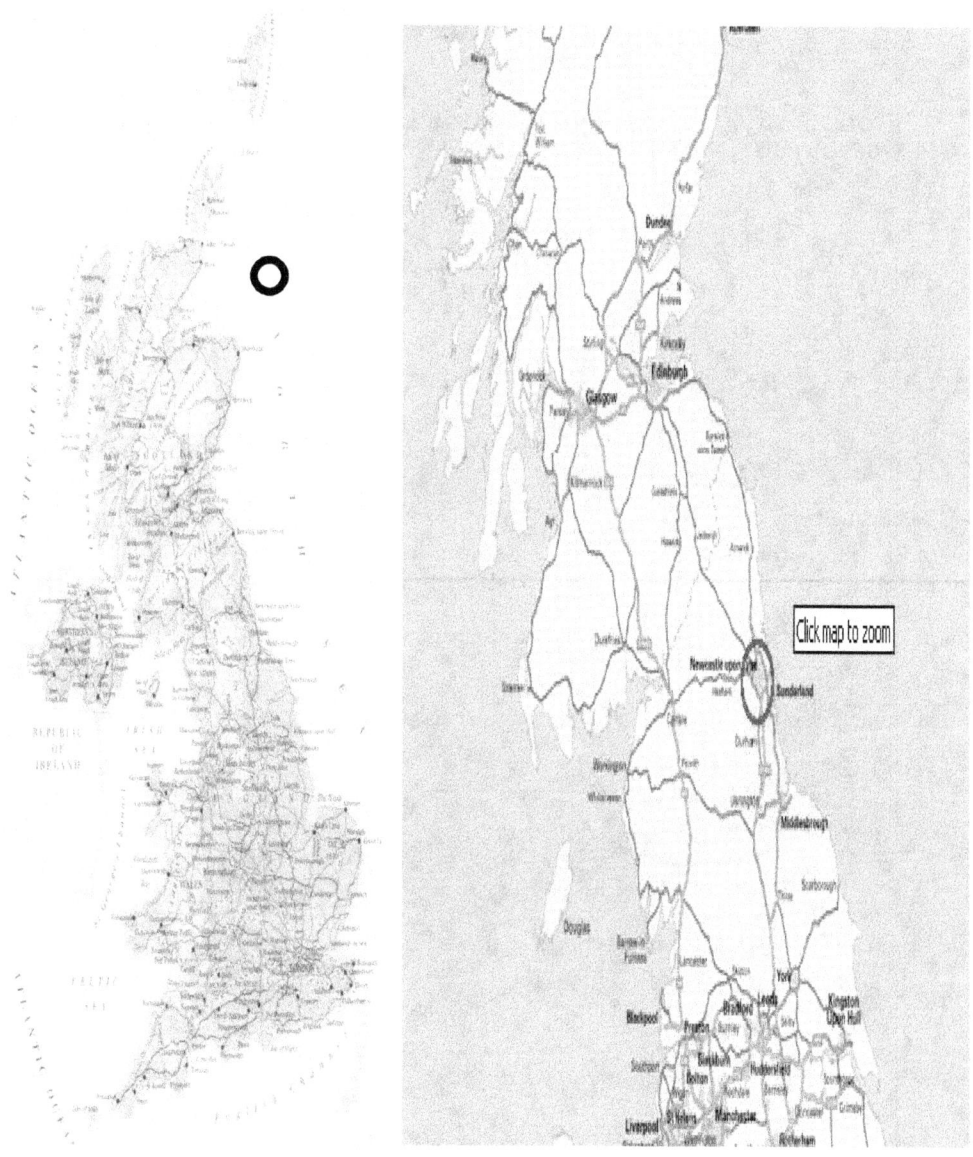

Newcastle upon Tyne and the

North East of England

Productivity (GVA per job) vs. Participation (jobs per population of working age) - 2003

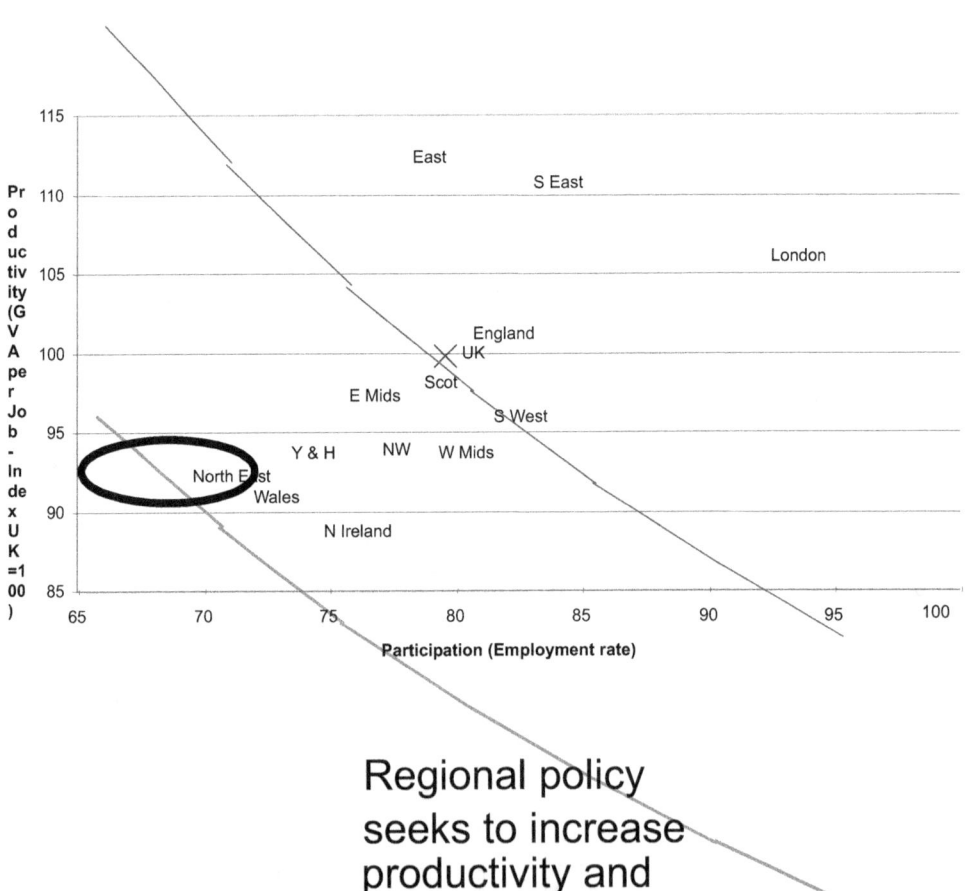

Regional policy seeks to increase productivity and participation

NE sectoral performance, sector growth and regional significance

	Sectoral Performance NE 2002 (GB = 100%)	GVA Growth GB 1992-2002	Regional LQ NE 2002
Agriculture, hunting, forestry & fishing	91%	-6%	59%
Mining and quarrying of energy producing materials + Other mining and quarrying	89%	2%	131%
Manufacturing	97%	26%	100%
Electricity, gas and water supply	87%	39%	142%
Construction	83%	58%	112%
Wholesale and retail trade (including motor trade)	86%	102%	88%
Hotels and restaurants	87%	126%	93%
Transport, storage and communication	89%	77%	89%
Financial intermediation	88%	107%	53%
Real estate, renting and business activities	84%	161%	69%
Public administration and defence[4]	76%	46%	121%
Education	101%	124%	132%
Health and social work	101%	110%	134%
Other services	78%	150%	89%

Source: ONS & ABI

Lower than average manufacturing sector performance

UK manufacturing has relatively low GVA growth

Increasing manufacturing productivity is a regional priority

North East Economy

Previously a strong reliance on traditional industries:

* Coal mining
* Ship building
* Power plant (steam turbines, switchgear etc.)
* Defense (tank factory)

All these sectors have been in long term decline.

Nissan Motors UK (Sunderland)

The Bluebird was the first UK Nissan car, which was produced in July 1986.
At the end of 2004, the plant produced 400,00 cars per year.

HENRY
FORD
MUSEUM
&
GREENFIELD
VILLAGE

From the Collections of
the Henry Ford Museum

Henry Ford's production
line was developed in
1913. The idea was
inspired by a trip to an
abattoir.

1909 Model T Ford

Any color you like provided it is Black! - Standardized

Any color you like provided it is black!

Model T Ford
1909

Scientific Management

"Whenever a workman proposes an improvement, it should be the policy of the management to make a careful analysis of the new method, and if necessary, conduct a series of experiments to determine accurately the relative merit of the new suggestion and of the old standard. And whenever the new method is found to be markedly superior to the old, it should be adopted as the standard for the whole establishment ", F.W.Taylor, Principles of Scientific Management, 1911.

Standardization and best practice deployment

Ford Mass Production System

- Minimized waste, maximized value
- Workers paid $5 per day, more than double the average
- Model T cars were cheap for customers, by 1918, half of all American cars were Model Ts.
- By 1927, 15,007,034 had been produced, a record which stood for the following 45 years.

Toyota Production System

- After World War II, Toyota was almost bankrupt.
- Post war demand was low and minimizing the cost per unit through economies of scale was inappropriate. This led to the development of demand-led pull systems.
- The Japanese could not afford the expensive mass production facilities of the type used in the USA so they instead focused on reducing waste and low cost automation.
- Likewise, Toyota could not afford to maintain high inventory levels.

Founders of the Toyota Production System (TPS)

Taiichi Ohno
(1912 †1990)

Shigeo Shingo
1909 †1990

Just-in-Time Manufacturing

"In the broad sense, an approach to achieving excellence in a manufacturing company based upon the continuing elimination of waste (waste being considered as those things which do not add value to the product). In the narrow sense, JIT refers to the movement of material at the necessary time. The implication is that each operation is closely synchronized with subsequent ones to make that possible" APICS Dictionary 1987.

JIT became part of Lean Manufacturing after the publication of Womack's
Machine that Changed the World in 1991

Faurecia, Washington, Tyne & Wear

Lean Manufacturing is a way of thinking

Lean Manufacturing goals

Lean Manufacturing

- Arose in Toyota Japan as the Toyota Production System
- Replacing complexity with simplicity
- A *philosophy,* a way of thinking
- A process of *continuous* improvement
- Emphasis on minimizing inventory
- Focuses on eliminating waste, that is anything that adds cost without adding value
- Often a pragmatic choice of techniques is used

Toyota Production System

- Technologies and practices can be copied.
- Most of the philosophies and techniques are widely disseminated.
- However, Toyota remains at the forefront, primarily because it is a learning organization.
- Problem solving methods are applied routinely and are completely ingrained.
- The employees are continually engaged in *Kaizen* (continuous improvement).
- Many aspects of TPS are based upon embedded tacit knowledge.

TPS: How the work is done

* Every activity is completely specified, then applied routinely and repetitively.

Because:

* All variation from best practice leads to poorer quality, lower productivity and higher costs.

* It hinders learning and improvement because variations hide the link between the process and the results.

It is necessary to make sure that the person performing the activity can perform it correctly and that the correct results are achieved.

7 Forms of Waste *'Muda'*

- *Overproduction* – most serious waste because it discourages the smooth flow of material and inhibits productivity and quality.
- *Waiting* – wastes time and money
- *Transport*
- *Inappropriate processing* – e.g. use of complex processes rather than simple ones. Over complexity encourages over production to try and recover the investment in over complex machines.
- *Unnecessary inventory* – increases lead-times and costs.
- *Unnecessary motion* – relates to poor ergonomics were operators have to stretch, strain etc. This makes them tired.
- *Defects* – physical waste. Regarded as an opportunity to improve. Defects are caused by poor processes.

Lean Manufacturing

* Philosophy
* Techniques – usually applied very pragmatically.

Lean Techniques

- Manufacturing techniques
- Production and material control
- Inter-company Lean
- Organization for change

Manufacturing Techniques

- *Gemba Kanri*
- Cellular manufacturing
- Set-up time
- reduction
  Smallest machine concept
- Fool proofing (*Pokayoke*)
- Pull scheduling
- Line stopping (*Jikoda*)
- I,U,W shaped material flow
- Housekeeping

'Genba Kanri' – Workplace Management

- System by which standards for running the day-to-day business are established, maintained controlled and improved
  .

Includes a number of methods:

- 5Ss
- Standard operations
- *Skill control*, including the assessment of individuals capabilities, the identification of job requirements, the development of a comparison matrix and the identification of training
- needs;
  Kaizen is a cost cutting approach that continuously makes small improvements to processes (Wikipedia, 2005)
- *Visual management*, the provision of notice boards for control information, stock, materials movement, health and safety and work methods.

5Ss Waller, D.L.,,1999,"Operations Management: A Supply Chain Approach", (Thompson, London)

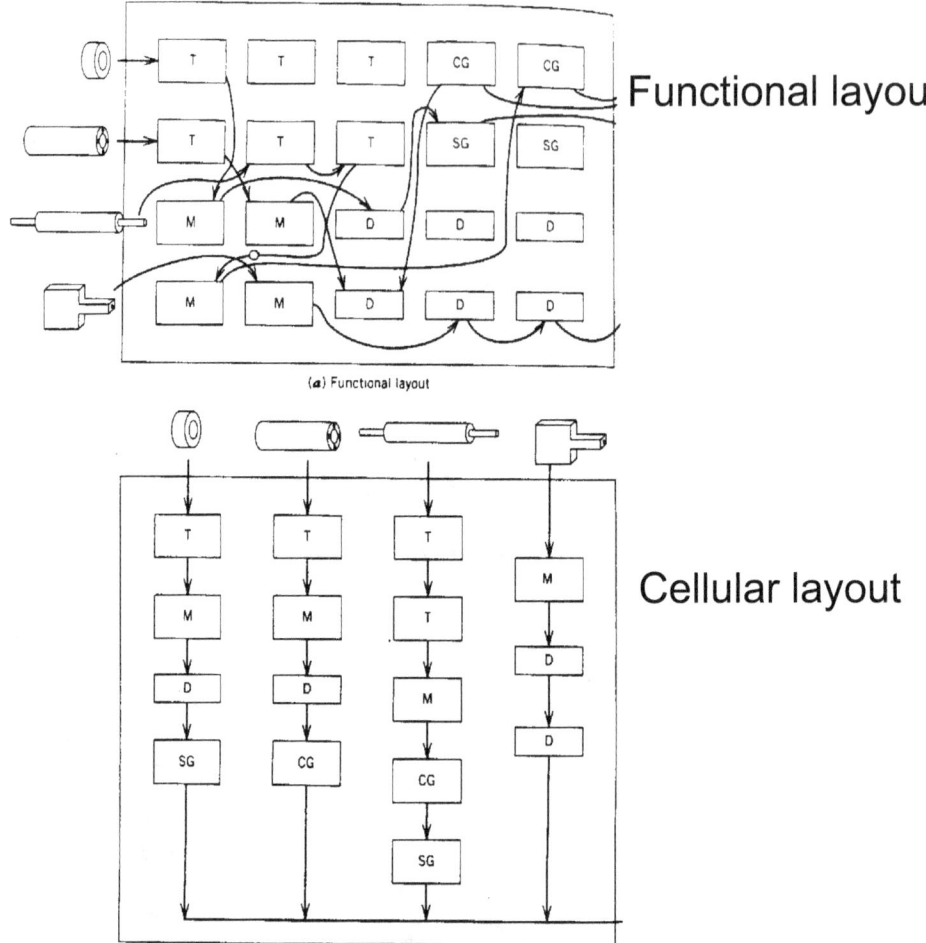

Functional layout

(a) Functional layout

Cellular layout

(b) Group layout (flow-line cell)

Functional layout

Power Generation Systems

Manufacturin
cells

A single machine acting as a cell

Multifunction double gantry mill

Power Generation Systems

Group Technology / Cellular Manufacturing

- Improved material flow
- Reduced queuing time
- Reduced inventory
- Improved use of space
- Improved team work
- Reduced waste
- Increased flexibility

Set-up Time Reduction

- Single minute exchange of dies (SMED) - all changeovers < 10 mins.

1. Separate internal set-up from external set-up. Internal set-up must have machine turned

2. off. Convert as many tasks as possible from being internal to external

3. Eliminate adjustment processes within set-up

4. Abolish set-up where feasible

Shingo, S. (1985),"*A Revolution in Manufacturing: the SMED System*", The Productivity Press, USA.

Set-up Analysis

- Video whole set-up operation. Use camera's time and date functions

- Ask operators to describe tasks. As group to share opinions about the operation.

Three Stages of SMED

1. *Separating internal and external set-up*

   doing obvious things like preparation and transport while the machine is running can save 30-50%.

2. *Converting internal set-up to external set-up*

3. *Streamlining all aspects of the set-up operation*

Single Minute Exchange of Dies (SMED)

Increases flexibility
Makes it easier to reduce
batch size
Reduces waste

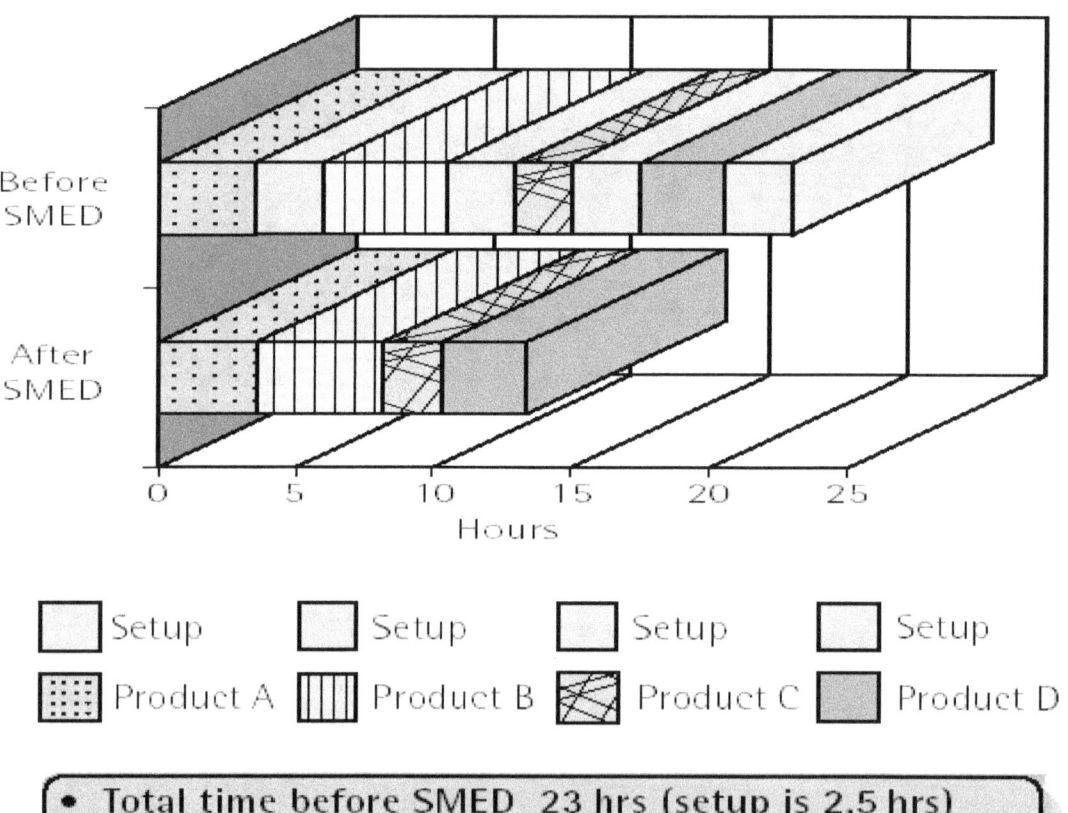

Legend:
- Setup
- Setup
- Setup
- Setup
- Product A
- Product B
- Product C
- Product D

- **Total time before SMED 23 hrs (setup is 2.5 hrs)**
- **Total time after SMED, 13 hr 32 min (setup is 8 min)**
- **Reduction of 41%**

Overall Equipment Effectiveness

- Open time – total time an operator available to work on a machine e.g. 8 hours per day
- Operator pause – coffee breaks, chatting, toilet breaks etc.
- Machine breakdowns
- Unplanned interruptions e.g. having to make modifications
- Machine set-up
- Low performance – throughput less than design.
- Scrap products

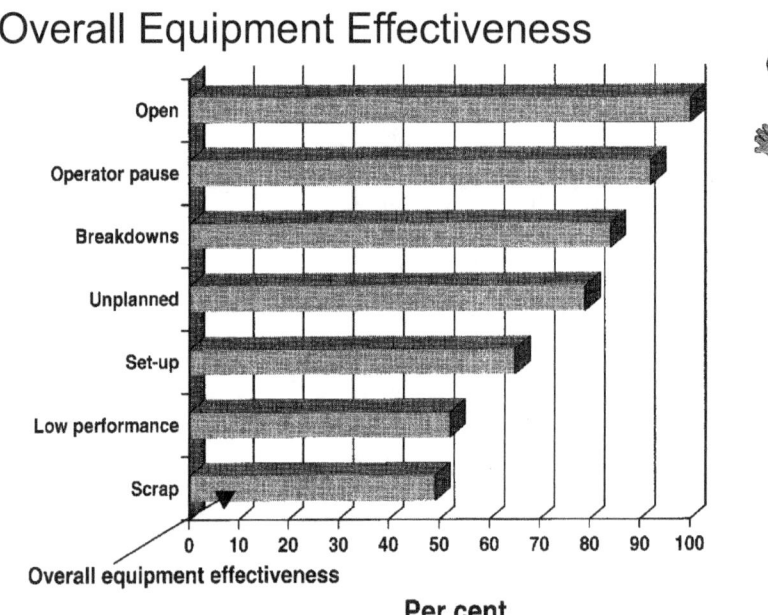

Overall Equipment Effectiveness

Small Machine Concept

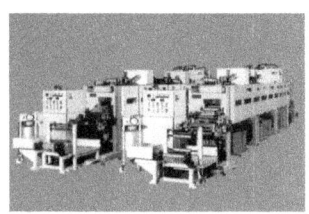 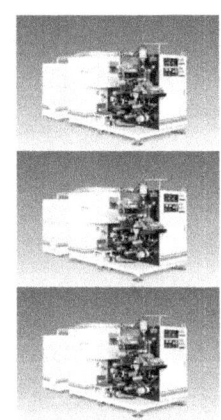

Using several small machines rather than one large one allows simultaneous processing, is more robust and is more flexible

Lean Material Control

- Pull scheduling
- Line balancing
- Schedule balance and smoothing (Heijunka)
- Under capacity scheduling
- Visible control
- Point of use delivery
- Small lot & batch sizes

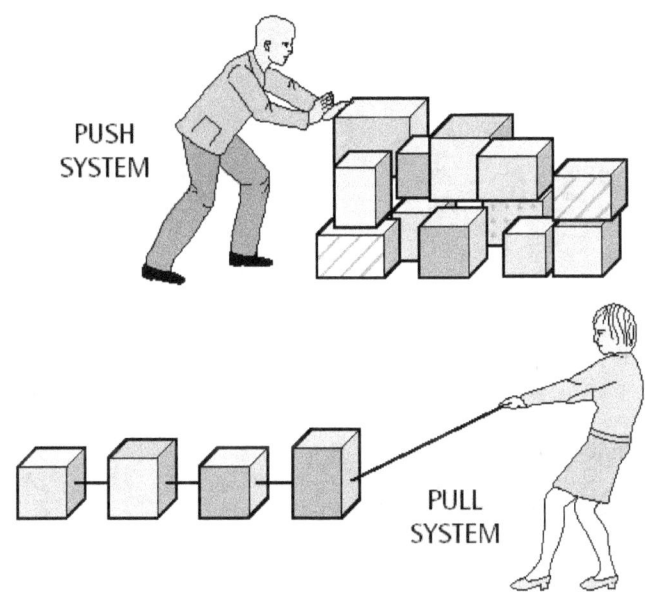

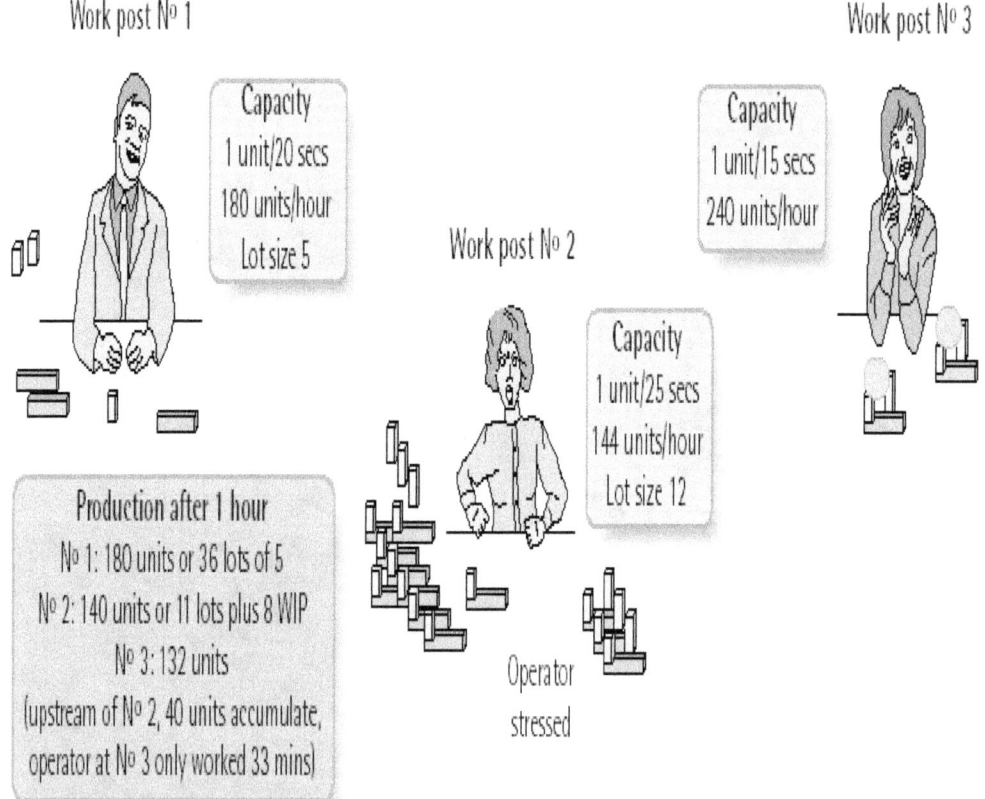

Operators work at their own pace trying to maximize output

Work post Nº 1

Capacity
1 unit/20 secs
180 units/hour
Lot size 5

Work post Nº 3

Capacity
1 unit/15 secs
240 units/hour

Work post Nº 2

Capacity
1 unit/25 secs
144 units/hour
Lot size 12

Production after 1 hour
Nº 1: 180 units or 36 lots of 5
Nº 2: 140 units or 11 lots plus 8 WIP
Nº 3: 132 units
(upstream of Nº 2, 40 units accumulate,
operator at Nº 3 only worked 33 mins)

Operator
stressed

Push system

Workers operate at their own pace trying to maximize output

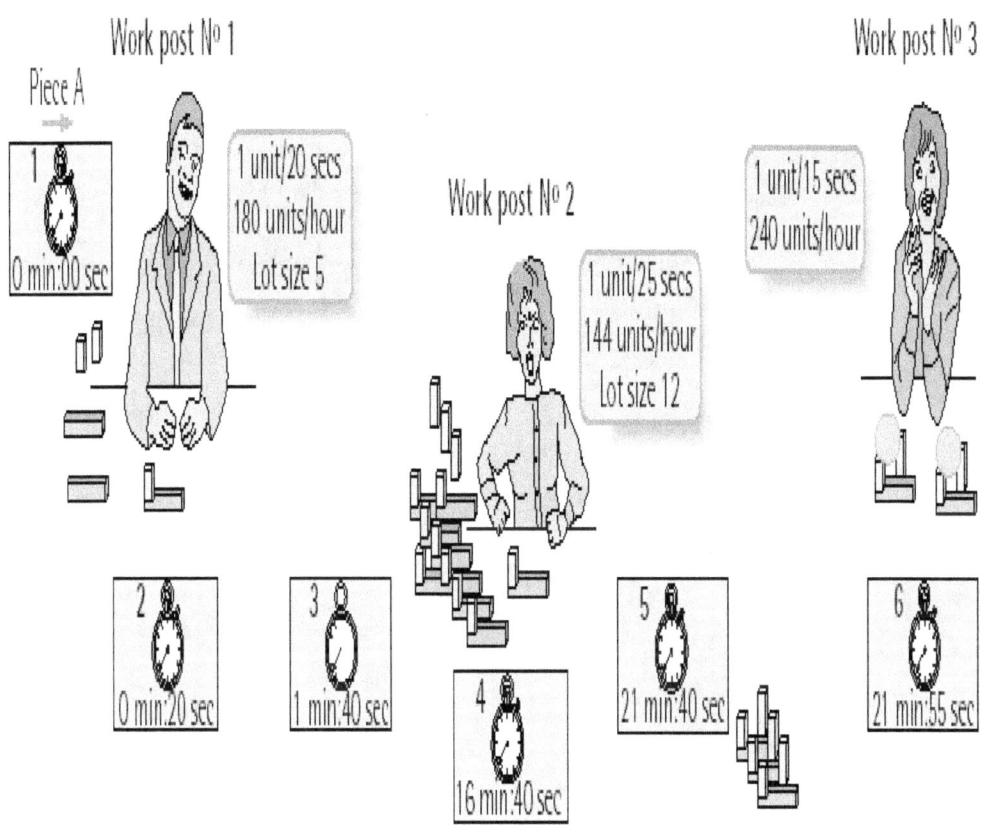

Operators work at their own pace trying to maximize output

Push system   **Lead time**

Operators wait if there is already one unit upstream at next post before producing

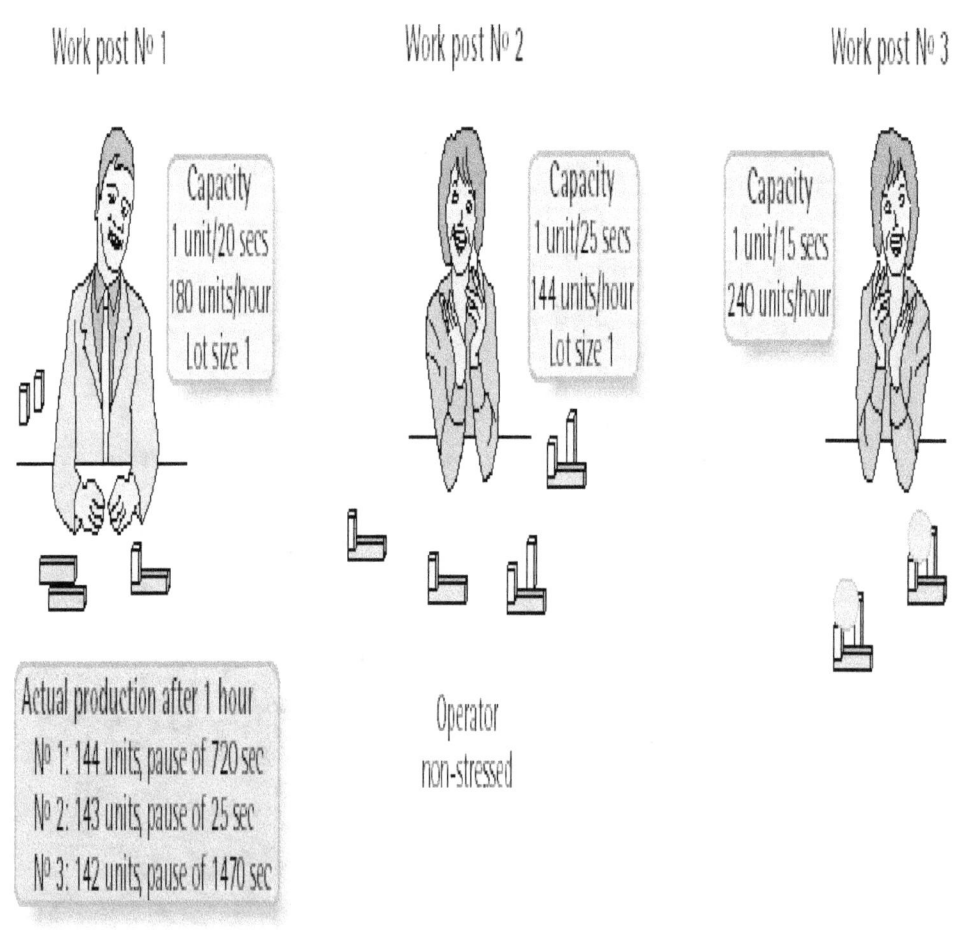

Work post № 1

Capacity
1 unit/20 secs
180 units/hour
Lot size 1

Work post № 2

Capacity
1 unit/25 secs
144 units/hour
Lot size 1

Work post № 3

Capacity
1 unit/15 secs
240 units/hour

Actual production after 1 hour
№ 1: 144 units, pause of 720 sec
№ 2: 143 units, pause of 25 sec
№ 3: 142 units, pause of 1470 sec

Operator
non-stressed

Pull system synchronized with demand. Lot size = 1

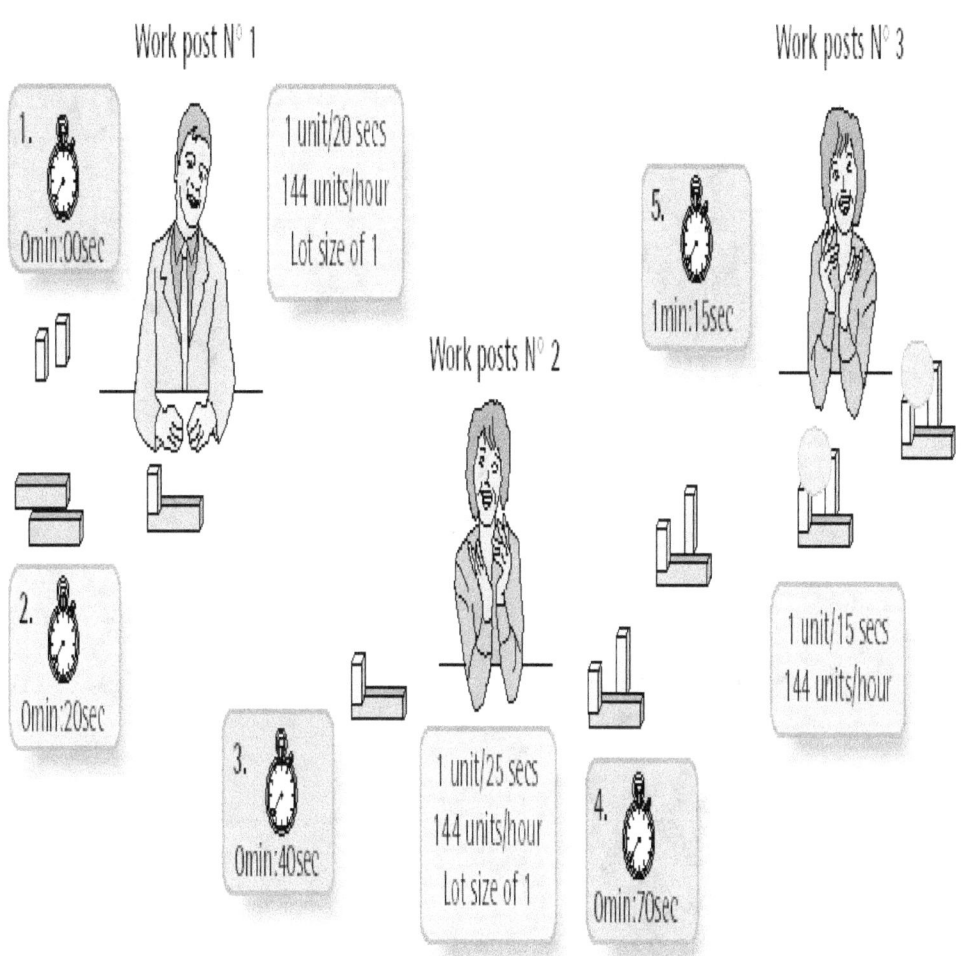

Upstream operator waits until downstream post needs a unit before producing

Pull system

Lead time

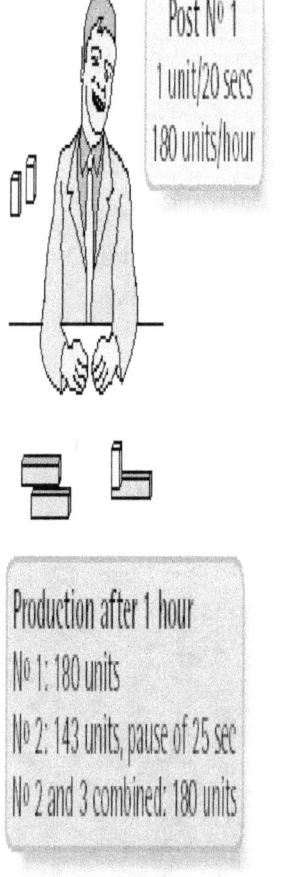

Work post № 1

Post № 1
1 unit/20 secs
180 units/hour

Production after 1 hour
№ 1: 180 units
№ 2: 143 units, pause of 25 sec
№ 2 and 3 combined: 180 units

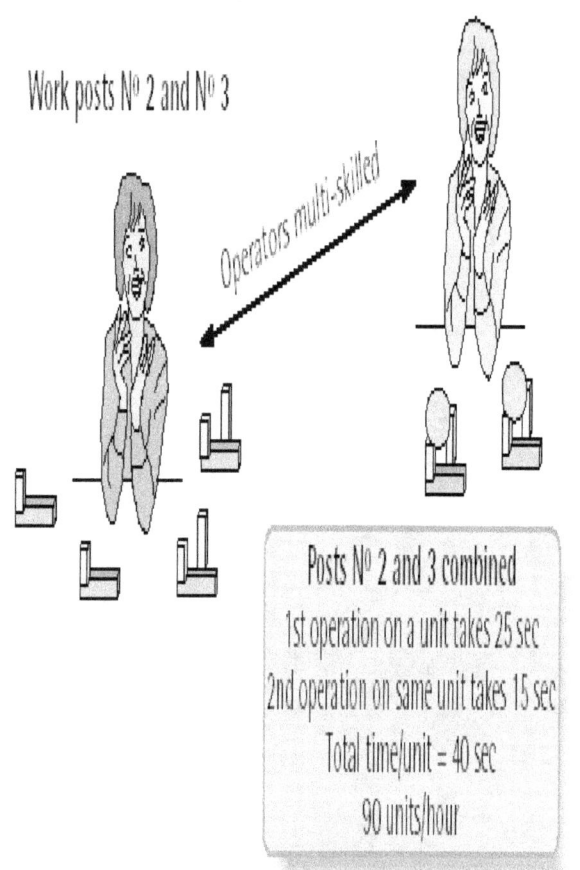

Work posts № 2 and № 3

Work posts № 2 and № 3

Operators multi-skilled

Posts № 2 and 3 combined
1st operation on a unit takes 25 sec
2nd operation on same unit takes 15 sec
Total time/unit = 40 sec
90 units/hour

Production after 1 hour:
WP1: 180
WP2&3 combined: 180
Increase = 36 per hour

Flexible workers in
Lean
combine WP2 & 3

"Pull" Systems

- Work centers only authorized to produce when it has been signaled that there is a need from a user / downstream department
- No resources kept busy just to increase utilization

Requires:

- Small lot-sizes
- Low inventory
- Fast throughput
- Guaranteed quality

Pull Systems

Implementations vary
- Visual / audio signal
- "Chalk" square
- One / two card Kanban

Lean Purchasing

- Lean purchasing requires predictable (usually synchronized) demand
- Single sourcing
- Supplier quality certification
- Point of use delivery
- Family of parts sourcing
- Frequent deliveries of small quantities
- Propagate Lean down supply chain, suppliers need flexibility
- Suppliers part of the process vs. adversarial relationships

Lean Purchasing

- Controls and reduces inventory
- Reduces space
- Reduces material handling
- Reduces waste
- Reduces obsolescence

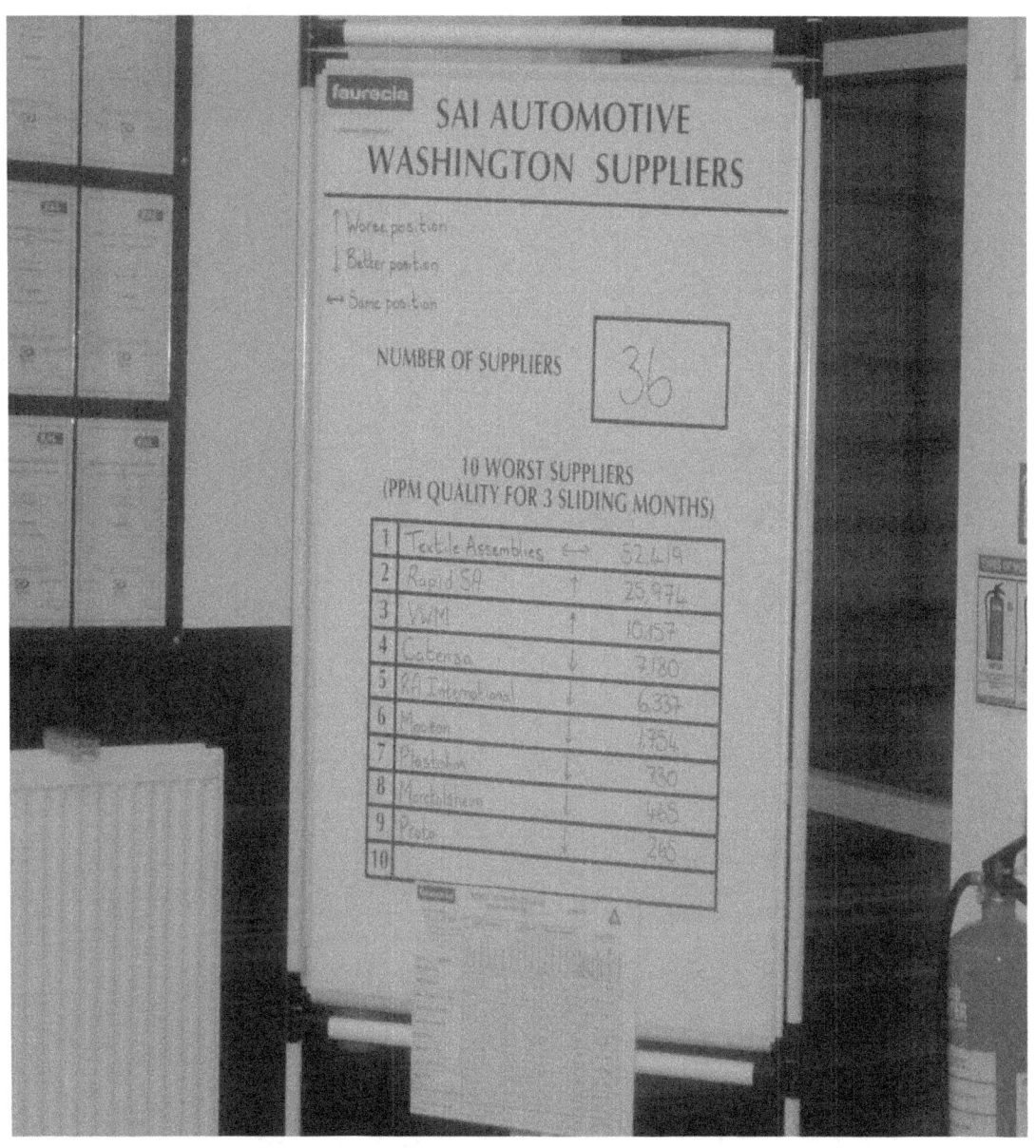

Notice placed prominently at the door at Faurecia

8	Marckolsheim		
9	Prota.	↓	265
10			

faurecia

WORST TEN SUPPLIERS FOR VENDOR RATING

June-04

Rank	Supplier	Quality (Demerits)	Cost (Score)	Delivery (Demerits)	Target 70% Overall Total
35th	Mouros				34.0
35th	Plastrom				58.6
34th	Textile Assemblies				62.8
33rd	Gefco/Transfesa				63.1
32nd	Marckolsheim				59.1
31st	Prota				65.2
30th	TRW France				76.5
29th	Coco Plastics				58.9
28th	RSM				71.2
27th	Faurecia Hambach Sist.				71.2
	All Supplier Average				71.8

Organization for Change

- Multi-skilled team working
- Quality Circles, Total Quality Management
- Philosophy of joint commitment
- Visible performance measurement
    - Statistical process control (SPC)
    - Team targets / performance measurement
- Enforced problem solving
- Continuous improvement

Total Quality Management (TQM)

* Focus on the customer and their requirements
* Right first time
* Competitive benchmarking
* Minimization of cost of quality
    * Prevention costs
    * Appraisal costs
    * Internal / external failure costs
    * Cost of exceeding customer requirements
* Founded on the principle that people want to own problems

The Deming Cycle

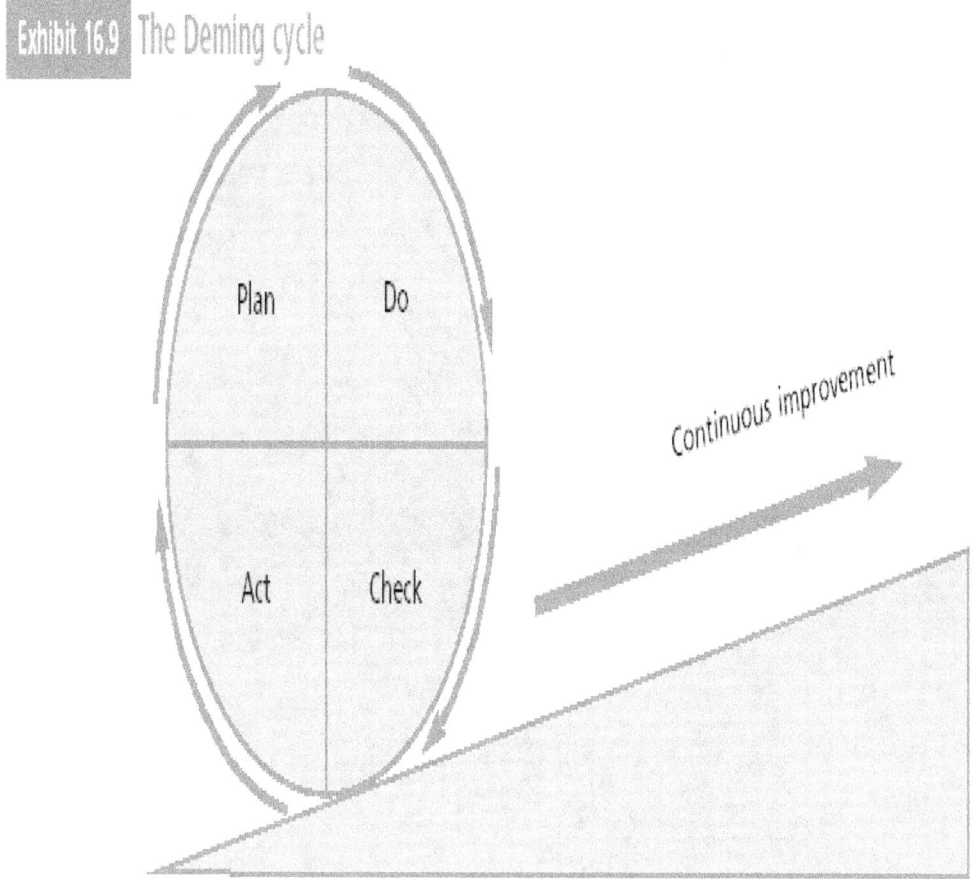

Exhibit 16.9 The Deming cycle

Plan · Do · Act · Check

Continuous improvement

ₙ𝑑 Edition", **Palgrave Macmillan**

Cause/effect (fishbone) diagram

Exhibit 16.13 An example of a cause and effect diagram

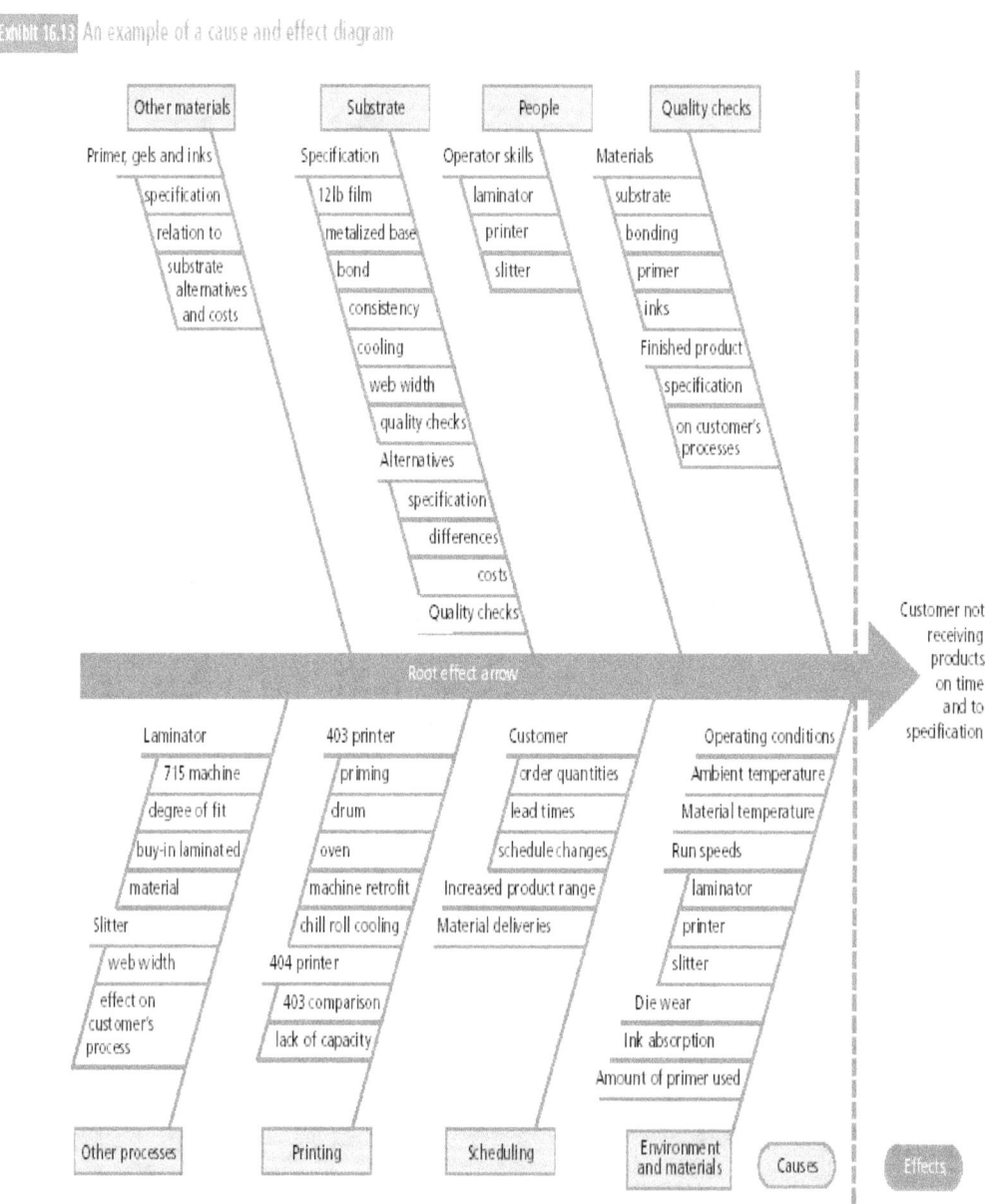

Lean Flexibility

- Set-up time reduction
- Small transfer batch sizes
- Small lot sizes
- Under capacity scheduling
- Often labor is the variable resource
- Smallest machine concept

Reducing Uncertainty

- Total Preventative Maintenance (TPM) / Total Productive Maintenance
- 100% quality
- Quality is part of the process - it can't be inspected in
- Stable and uniform schedules
- Supplier quality certification

Total Preventative Maintenance (TPM)

- Strategy to prevent equipment and facility downtime
- Planned schedule of maintenance checks
- Routine maintenance performed by the operator
- Maintenance departments train workers, perform maintenance audits and undertake more complicated work.

The problem with inventory

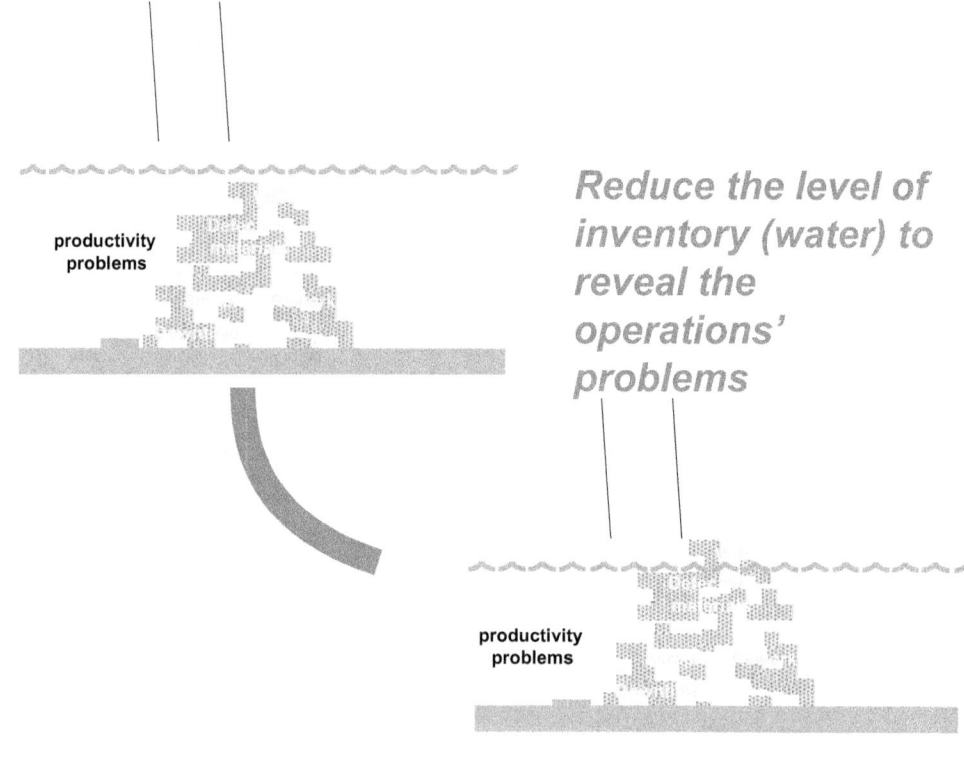

productivity
problems

*Reduce the level of
inventory (water) to
reveal the
operations'
problems*

productivity
problems

Operational prerequisites

- Level schedules
- Frozen schedules
- Fixed routings
- Frequent set ups
- Small and fixed order quantities
- High quality conformance
- Low process breakdowns
- Labor utilization not the key factor
- Employee involvement

Operational prerequisites

- Level schedules
- Frozen schedules
- Fixed routings
- Frequent set ups
- Small and fixed order quantities
- High quality conformance
- Low process breakdowns
- Labor utilization not the key factor
- Employee involvement